H. L. (Harry Luman) Russell

Bacteria in their Relation to Vegetable Tissue

H. L. (Harry Luman) Russell

Bacteria in their Relation to Vegetable Tissue

ISBN/EAN: 9783337069643

Printed in Europe, USA, Canada, Australia, Japan

Cover: Foto ©berggeist007 / pixelio.de

More available books at **www.hansebooks.com**

IN THEIR

RELATION TO VEGETABLE TISSUE

A DISSERTATION

PRESENTED TO THE BOARD OF UNIVERSITY STUDIES OF THE
JOHNS HOPKINS UNIVERSITY FOR THE DEGREE
OF DOCTOR OF PHILOSOPHY

BY

H. L. RUSSELL

1892

PRESS OF
THE FRIEDENWALD COMPANY
BALTIMORE

Contents.

BACTERIA IN THEIR RELATION TO VEGETABLE TISSUE.

By H. L. RUSSELL.

The relations of that group of micro-organisms known as bacteria to the animal kingdom have within the past two decades been made the subject of unusual attention.

No department of the rapidly developing branches of science has been the field of greater activity, yet, strange to say, the relations which these organisms bear to the co-ordinate branch of biology, the vegetable kingdom, have been greatly neglected.

Botanists have studied bacteria as a class, more or less, in order to determine, if possible, their affinities with other low forms of life, but the greater activity in this field has been largely due to the close relation which they bear to medicine in the etiology of disease.

In regard to the relations which they bear to and the influence that they exert upon higher plant life, the data we possess are meager. The progress which has already been made in this department of plant pathology comes largely from this side of the Atlantic, and to a prominent American botanist, Prof. T. J. Burrill, belongs the honor of having been the first to work out the causal relation between a specific microbe and, a plant malady (pear-blight, 1878).

Although Prof. Burrill's work was done over a decade ago, comparatively little notice has been taken of it by European writers, with but few exceptions, and the majority of the text-books that refer to it at all regard the case as not thoroughly proven; but upon what grounds these conclusions are based it is quite impossible for one to understand who has had access to the already widely published data.

It seems to be a wide-spread belief that plants do not suffer to any great extent from the attacks of these micro-organisms. The reasons usually assigned for this so-called immunity are various. Chief among them, however, is that which bases the freedom of plants from bacterial attack upon the acidity of the plant-tissue. Other subsidiary reasons are also advanced by various authors.

Flügge[1] states that bacteria almost never attack higher plants, giving as an only exception Wakker's hyacinth-disease. The low temperature of plants and the chemical composition of vegetable juices he regards as very unfavorable for the development of bacteria, more especially as the cell-juices almost always possess a distinct acid reaction, and thus protect the plant against these micro-organisms, which are so sensitive in this respect.

Hartig, in the recent edition of his Lehrbuch der Baumkrankheiten (p. 37), also urges the view that the acid reaction of plants prevents the growth and development of bacteria, and that they play a very unimportant rôle in the production of plant-disease. In the hyacinth-disease above referred to, he says the bacteria do not attack sound, well-ripened bulbs under normal conditions,but only when the tissues have been more or less injured by wounds or previous attacks of fungi.

He notices, also, Burrill's claim that the pear-blight is caused by a specific bacillus, but is somewhat skeptical that the form referred to is anything but a secondary accompaniment of the malady.

DeBary is inclined to support the general views advanced by Hartig, but in his last edition of Die Bakterien (S. 36) he states that it might be possible for bacteria to gain access, through stomata, into the tissues of higher plants, but that this is probable is yet undetermined and needs further investigation.

These observers, all of them recognized authorities in the realm of pathology, seem to regard it as quite improbable that bacteria have any important bearing upon the production of plant-disease. Whether this unanimity of expression is due to the actual absence of bacterial plant-maladies in Europe generally, or because investigations have not been directed in these channels, can only be inferred.

In consideration of the fact that this branch of vegetable pathology is of increasing importance, and that the reasons assigned for the apparent exemption of plant-tissues from the attacks of micro-organisms have been largely based upon the general law known in regard to bacterial life in general, it was deemed advisable that a series of investigations should be carried out with different micro-organisms, to see what effect contact with the living plant-tissues would have upon them; so, at the suggestion of Prof. Welch, this topic was taken up for consideration.

[1] Die Mikroorganismen, S. 515.

I have been unable to find any literature upon this particular question with the exception of a preliminary report by Lominsky, who worked mainly with those forms which are pathogenic for animals. His original paper is in Russian, so I have been forced to rely solely upon an abstract (Cent. für Bakt., Bd. VIII, 325) for his data.

Aside from this single exception, I find no general series of experiments recorded as giving the effect of vegetable tissues upon various forms of bacterial life.

METHOD OF EXPERIMENT.

The following outline will indicate the manner in which the experiments were carried out. Fresh cultures of the various microorganisms were always taken (usually bouillon cultures 12–24 hrs. old), so as to insure the introduction of non-sporogenous material. A young growing stem was selected, so as to give the most favorable conditions for the development of the organism. It was first washed with sterile water and then pierced with a fine sterilized platinum needle. Into this minute opening a tiny droplet of culture fluid was injected from a capillary pipette. The slight puncture caused by the penetration of the needle was then closed from the influence of air and possibility of accidental contamination, by sterile vaseline.

The results were determined by excising a section of the infected stem, the surface having been slightly flamed in a Bunsen flame.

The cortical layer was then removed with a sterile scalpel, leaving the inner tissue into which the organism had been injected. Quite thin sections of this remaining tissue were cut, under aseptic precautions, and inoculated into tubes of melted gelatine and roll cultures made therefrom. As the fluid gelatine easily penetrates the plant-tissue, a moderately thin section may be examined, under considerable magnification, with ease. The tissue was sectioned, not only at the inoculation point, but at varying distances above and below. By growing these serial sections in sets of culture tubes, one is able to determine how far the bacteria have spread throughout the plant.

An objection to this method lies in the fact that small variations in the germ-content cannot be detected, as it is quite impossible to prepare the tissue so that all germs present can develop; but where cultures made from tissue taken at the point of inoculation reveal but few germs, we may safely conclude that they have either been killed off by the plant or died from insufficient nutrition.

On the other hand, an increase can only be considered probable where the cultures, not only from the tissue immediately surrounding the inoculation point, but at a distance from it, reveal a large number of germs. Even the fact that bacteria are to be found at a greater or less distance from point of introduction does not necessarily show that an *actual* increase has taken place. Their presence at this point might be considered as due either to simple diffusion or to mechanical transportation by the fluids of the plant. The effect of these possible factors will, however, be shown later to be quite nugatory.

If macroscopical changes are to be seen in the tissue, it would be of itself sufficient evidence that actual multiplication of the microorganisms had taken place. In addition to culture methods to determine the presence or absence of bacteria in the infected tissues, sections were also subjected to microscopical examination.

But this method proved quite unsatisfactory, except where the bacteria in the tissues were numerous, as in the case of actual infection. The granular detritus and peculiar rod-like masses of protoplasm often found in matured cells make it extremely difficult to differentiate the bacteria in an unstained condition. The use of staining methods, so successful in the differentiation of bacteria in sections of animal tissue, have not as yet been successfully applied to bacteria in plants. The aniline stains, toward which the bacteria are so susceptible, seem to impregnate the vegetable cell and its membranes with great ease, and in the use of decolorizing agents, parts of the plant cell retain the stain as deeply as do the bacteria.

Before detailing the results of the experiments made, we will consider briefly the presence of bacteria in normal uninjured plant-tissues. This, for a considerable time, has been a debatable question, and the recorded results of numerous observers are somewhat at variance with one another.[1] The preponderance of evidence is, however, certainly against the view that micro-organisms are present normally in the tissues of higher plants, and this conclusion harmonizes well with what we know in the domain of animal life.

In the examination of plant-tissues I have made a large number of cultures, according to the method described above, from plants selected as healthy in all respects, without being able to isolate bacteria from them. Bacteria, however, were often found in tissue which had been wounded from any cause, and in some cases in such numbers as to lead one to think that they had possibly multiplied in the plant-tissue. This can happen from the local death of the wounded tissue, which will enable the micro-organisms to gain a foothold, and even though they may not be able to *grow* within the living plant, they are able to *exist* for a considerable length of time (as will be shown by the results of artificial inoculation), and thus come to be enclosed in the plant by the healing over of the wounded tissue. This is, I think, a probable explanation of the data recorded by some observers who claim to have actually isolated saprophytic forms from plant-tissues.[1] Lesions so slight as to escape notice, especially in root crops, would allow the access of saprophytic forms to the tissues of the plant, where they might survive for a considerable length of time.

From the results of my own experiments, the conclusion seems evident that, *normally, the healthy plant, with intact outer membranes, is free from bacteria within its tissues.*

In the tabulated results obtained by the artificial inoculation of different bacterial species into vegetable tissue, they will be classified according to their nutritive adaptation.

TABLE SHOWING ACTION OF SAPROPHYTES IN PLANT TISSUE.

Name of Germ.	Date of Inoculation.	Date of Close of Exp.	Period of Incubation. Days.	Host Plant.	Result.
B. prodigiosus............	X. 20	XI. 17	27	Tradescantia.	** [2] ✓
"	X. 20	II. 1	103	"	—
"	XI. 26	XII. 5	10	Geranium.	*
"	XII. 20	II. 2	42	"	**
B. butyricus...).........	XII. 20	II. 2	42	"	**
"	XI. 28	XII. 10	13	Lima Bean.	*
B. luteus................	XII. 20	I. 28	40	Geranium.	**
B. megaterium...........	XI. 19	XI. 30	11	Lima Bean.	*
"	XI. 19	XI. 30	11	"	—
"	I. 12	II. 25	44	Geranium.	*
B. coli commune..........	XII. 1	XII. 20	19	"	**
" "	XII. 1	XII. 30	29	"	**
B. ac. lactici.............	I. 12	II. 16	35	"	**
B. fluorescens	I. 12	II. 24	43	"	**
B. lactis aerogenes........	I. 4	II. 14	·10	"	*

[1] Fazio and others: Revista Internaz. d'Igiene, (1890), I, 3.
[2] Explanation of signs : * denotes presence in moderate numbers.
        ** " " " large "
        — " absence in culture entirely.

The above table indicates that a number of different forms which are ordinarily saprophytic in their method of nutrition, are able to exist within the plant for a considerable period of time, and in some cases show evidence of a considerable increase. This multiplication does not, however, reach a stage macroscopically observable. There is usually a slight "browning" or discoloration of the tissue at the seat of inoculation, but this is due to the slight injury caused by the inoculating needle, even though the opening is protected from the influence of the air by vaseline.

The results which were obtained by Lominsky[1] in regard to the growth of Bac. prodigiosus I was unable to confirm. He states that this germ, inoculated into the leaves of certain plants, produced brick-red spots and stripes which were to be seen by the naked eye. In the above experiment bacteria were demonstrated as present in large numbers in tissue even after a considerable lapse of time, but no signs could be detected of a change in the cellular structure in any case.

In regard to forms which are naturally pathogenic for animals, we might expect *a priori* that they would be unable to survive for such a length of time, or show as marked an increase, as saprophytic forms, which are, as a rule, less sensitive in regard to the substratum for their development. This expectation was realized, as will be seen from the subjoined table, which comprises those forms that are natural facultative parasites on animals.

In this group of parasites I find, with but few exceptions, that they are unable to compete with the unfavorable environment to which they are subjected in the plant. The large majority of them are not able even to hold their own, but gradually succumb to their unfavorable surroundings. Here again I failed to verify some of the results obtained by Lominsky. His experiments were confined to the action of plant-tissues upon anthrax, the typhoid bacillus, and Staph. pyog. aureus. He made quite a number of experiments, and found that both the anthrax and the pyogenic organism increased and were able to form colonies in the tissues. Anthrax grew rapidly for a time, formed spores, and finally seemed to undergo degeneration. The typhoid-fever bacillus was unable, however, to live beyond a few days, and even then showed degenerating peculiarities.

---

[1] *Loc. cit.*

TABLE SHOWING ACTION OF ANIMAL PARASITIC FORMS IN VEGETABLE TISSUE.

Name of Germ.	Date of Inoculation.	End of Experiment.	Period of Incubation. Days.	Host Plant.	Result.
B. pyocyaneus	XI. 27	II. 4	69	Begonia. cult.	**
"	XI. 28	XII. 30	32	Geranium.	**
"	XI. 27	I. 2	36	Penthorum.	**
B. anthracis	XI. 20	I. 26	38	Gerauium.	—
"	XI. 19	XI. 30	11	Lima Bean.	(6)[1]
"	XI. 20	XI. 25	5	Echino cactus.	(2)
Staph. epid. alb.....	XI. 20	I. 28	40	Geranium.	—
Staph. pyog. aureus.	I. 12	II. 23	42	"	—
" " "	XII. 10	XII. 23	13	Lima Bean.	(3)
Mic. cereus flav......	I. 12	II. 19	38	Geranium.	(4)
Cholera gallinarum..	II. 20	III. 10	18	"	*
Schweineseuche.....	III. 8	III. 25	17	"	**
Mic. tetragenus......	III. 22	IV. 16	25	"	—
Bac. diphtheriæ.....	III. 8	III. 18	10	"	—

[1] Numbers in parenthesis indicate number of colonies found in culture made from infected tissue.

It is noteworthy in the above table that the pyogenic organisms in general do not seem to be especially resistant. With the single exception of the blue pus-germ, they succumbed to the unfavorable influence of the plant-tissues.

The consideration of the third class, that of bacterial plant parasites, brings us to those forms which are, in a restricted sense at least, the natural enemies of vegetable life.

I have found it impossible to obtain cultures of more than a few of the germs which have been reported as having been isolated in the various plant-maladies, as in many cases cultures are not kept in stock, even by the discoverers of the germ.

Of those secured I made a series of infection experiments in a number of different hosts, to ascertain the effect of vegetable tissues in other than their natural hosts.

The pear-blight germ grown in a Begonia-plant for 30 days showed at end of that time large numbers at inoculation point, but not distributed throughout the plant. The same result was found when injected into Phaseolus vulgaris for 30 days, also in Ph. lunatus for 16 days. In Tradescantia alba, no trace could be found at the end of 60 days' incubation in this tissue. Bac. avenæ was injected into tissue of Begonia, onion, corn, wheat, and squash, but in no case was any pathological change macroscopically observable. The bacilli were not killed out in the plant-tissue, however, as they were isolated from Begonia and squash in large numbers, after 30 days' incubation in these tissues, but their presence was confined to the tissue contiguous to point of introduction.

The results of the foregoing inoculation experiments made with various forms of micro-organisms, saprophytes as well as parasites (both for animals and vegetables), show that these germs in many cases are able to live in the plant-tissues for a considerable length of time. A number of the different forms, particularly saprophytes, are able to grow and spread throughout the plant to a limited extent. Of the parasitic species tested, very few showed any tendency to thus spread. Even those forms that are natural parasites of certain higher vegetable species showed no power to spread in plants which were not their natural hosts, but they were able to live at inoculation-point for a considerable time.

The possible objection, already alluded to, that the distribution of the bacteria, which was noted in many cases, may not indicate actual growth, will now be considered.

The observed facts are these: The distribution of the micro-organisms in the plant-axis, as determined by culture experiments, always took place in an ascending direction. This distance varied from 30–50 mm. from point of introduction, but in no case were bacteria found more than 2–3 mm. below inoculation-point.

Germ.	Culture from Inoc. Point showed :	Culture made from Tissue taken.
Bac. luteus in Geranium 40 days....	1850 colonies,	10 mm. above, 1764
B. fluorescens " 43 "	4200 "	5 " " 3850
" " " "	" "	3 " below, 350
B. butyricus " 48 "	104 "	5 " above, 45
" " " "	" "	10 " " 20
B. acidi lactici " 35 "	6500 "	5 " " 4200
" " " " "	" "	25 " " 2250
" " " " "	" "	3 " below, 2000

Of course the numbers given above do not represent the total number of bacteria present, owing to the difficulty of preparing tissue so that *all* can develop. Besides this fact, the developing colonies were observed to be usually *intracellular* and not in the spaces of the plant.

Now let us consider the two possible theories, besides that of actual growth, which suggest themselves.

First, that of diffusion. For this a fluid substance is necessary that is continuous throughout the plant. As the cellulose wall and ectoplasm of the vegetable cell act as an effectual filter of solid particles, there would be no chance for direct diffusion from cell to cell. The inability to utilize the intercellular spaces for this purpose is equally evident, for these are, under normal conditions, filled only with a saturated vapor, and not fluid substances, and therefore unable to function as a means of diffusion.

If simple diffusion were operative, then, too, we would expect to find the bacteria diffused as far below the point of introduction as above, especially as gravity would aid in this result. This is contrary to the experimental facts.

Now, is it possible to explain this distribution as a result of the transpiration currents in the stem?

Whatever may be the ultimate outcome of the conflicting theories regarding the *locus* of the transpiration stream, it rests upon the imbibitory and osmotic powers of certain vegetable cells. But this stream can only carry substances in solution through these vegetable membranes, and therefore could not function as a transporter of solid bodies like bacteria. This was demonstrated by cutting the stem of a thrifty growing plant *under water*, and then transferring it to a vessel containing a nutrient solution to which a dilute culture of a germ had been added. It was found that in the exercise of the ordinary processes of vegetation, no germs were detected in the tissue to any considerable distance above the water level.[1]

Then, too, the distribution of the bacteria in different tissues, such as the cortical and pith parenchyma, as well as the fibro-vascular tissue, could hardly be explained by the action of this current.

If the germs were mechanically transported, why was it that only certain forms, notably saprophytes, were selected? This cannot be accounted for on the ground of size or independent motility of the organisms. Some forms, such as B. amylovorus, were able to exist in large numbers at inoculation-point for a considerable length of time,

---

[1] It is possible, however, that a slight amount of fluid may be drawn up a short distance into the intercellular spaces and vascular lumina, in order to equalize the negative pressure of the contained air which is found often in these cavities.

but did not seem to be able to spread throughout plants which were not its natural host.

From the above considerations it will be seen that there is no reasonable ground to support the hypothesis that the bacterial distribution is purely physical.

As no actual openings are known to exist in the walls of these cells (if we except the very minute pores through which the plasmic strands pass), it is difficult to understand how the bacteria are able thus to spread from cell to cell unless they possess the power of penetrating the cell wall by the action of vital forces.

This ability would require physiological activity which could not be present unless they were able to exercise their ordinary metabolic functions.

This penetrative power, among certain forms, is noteworthy when we compare it with the results obtained by injecting different species into the animal body.

Von Fodor[1] found that when B. termo, B. subtilis and B. megaterium were introduced in large numbers into the jugular vein of a living rabbit, they disappeared completely after a lapse of four hours.

Wyssokowitsch[2] also determined that the time necessary to completely destroy all bacteria contained in one c.c. of culture fluid, when introduced into the animal body, varied from 15 min. with Spir. tyrogenum to 7 hours with B. acidi lactici.

The rapid disposal of these forms in the animal body is correlated, as Nuttall[3] has shown, with the germicidal property of the blood-serum. The action of plant-tissue upon bacteria is in no case comparable to this, and would suggest that the plant is not protected in a similar manner. This point will, however, be considered in detail under another head.

The possible explanation for the *upward* distribution of germs may rest upon the principle that growth always follows the lines of least resistance. Not only are food materials more abundant in the rapidly growing apex, but the thinner and less developed cellulose walls offer much less resistance to the spread of the germs than the more matured cell-membranes of the older tissue.

---

[1] Von Fodor : Arch. f. Hyg. IV (1886), 129.
[2] Wyssokowitsch : Zeit. f. Hyg. I (1886), 3.
[3] Nuttall: Zeit. f. Hyg. IV (1888), 353.

Can bacteria penetrate the intact healthy tissue of the plant?

In considering this question we may disregard, in this connection, those cases where bacteria have gained access to the inner tissues by means of wounds, and have been able to live there for a certain time. The question as stated above is of practical importance in pathology, for if micro-organisms are able to penetrate the *intact tissues*, this will explain the way in which infectious material may be distributed from plant to plant.

The epidermal tissues of the plant are much more resistant to external influences than the parenchymatous elements. This is due to an outer layer of pure cutin, or to the impregnation of the cellulose walls with cutinized layers. This resistant sheath is replaced in the older plant by a thicker and more resistant corky layer. These protecting tissues are not perfectly continuous over the exterior of the plant, but are pierced by numerous small openings, the stomata, which afford a direct communication between the surrounding atmosphere and the inner cells of the plant.

What is now to prevent the entrance of micro-organisms through these minute openings in the outer membrane?

As regards fungi we know that some species, such as Cystopus candidus, the common white rust of Cruciferæ, do gain access to the inner tissues, first, by sending their germ-tubes through the stomata into the intercellular spaces. As these spaces are devoid of nutrient material, they must offer but poor conditions for growth to any organism that is not able to extract its nutriment from the living cell, either by haustoria or by penetrating the wall and, by means of ferment action, obtaining access to the protoplasmic materials of the cell.

With bacteria that are not adapted for a parasitic existence in plant-tissue it is not yet definitely determined whether they can enter by means of these natural openings. I was unable to isolate from the tissue of different plants any bacteria, although the pots and their plants were watered for several days with dilute infusions of the different germs. The results I obtained are not at all in harmony with those of Lominsky, who found that wheat could infect itself naturally in soil seeded with different species of bacteria. Not only was he able to isolate from the roots of the growing plant all the species which he added to the soil, but he found them also *in* the tissues of the stem and leaves. This result would have been much more convincing had he used larger plants than wheat, as it is quite possible that the bacteria isolated came from the surface rather than the inner tissues of the plant.

Although it is extremely doubtful if those micro-organisms whose mode of nutrition is not adapted for parasitical existence in vegetable tissues, can enter the plant without the intervention of an actual wound, it is much more probable that those forms naturally parasitic on plants may sometimes succeed in thus getting a foothold in the tissues.

Bolley,[1] in his work on surface scab of potatoes, tried the experiment of infecting sound, healthy, growing tubers with liquid cultures of the scab bacilli. The growing tubers, after thorough cleansing, were immersed in a fluid containing an infusion of the scab bacilli, and the glass vessel was then protected from outside contamination. In thirteen days the tubers so treated had decayed, while a control test with sterilized water showed tubers perfectly sound and the water clear. He reasons from this that the bacteria penetrated the growing lenticels of the tuber. He also repeated the experiment by saturating with an infusion of scab bacilli sterilized soil in which he transplanted the growing tubers. Control tests, watered with distilled water instead of the diluted culture, were made, and he found that infection took place when the bacteria came in contact with the healthy tuber.

With the pear-blight germ I have found it impossible to infect even the young budding leaves and stems of the pear, when these organs were several times sprayed with a culture of the germ. Susceptible as well as refractory varieties failed to succumb to the disease when subjected to this manner of infection. Atomizing the flower clusters, however, usually yields positive results, according to Waite. Here the tissue is thinner walled, and in some places, as the nectary, is destitute of cutin, so that the bacteria have less difficulty in effecting an entrance. The highly nutritious nectar affords them an excellent medium for growth, and here they are able to thrive until, as Waite has suggested, they get a foothold. It is quite possible that this intermediate stage of development affords an opportunity for the accumulation of a ferment by which the germs are able to more easily penetrate the subjacent tissue.

[1] Bolley: Agric. Science, Vol. IV, 250.

Infection experiments with Galloway's oat-disease succeeded usually with young plants when they were simply sprayed, but older and more developed plants failed to "take" the infection this way. This may possibly indicate that the stomata do not function as a means of entrance, as the older plants are furnished with these structures as well as the young seedlings.

Kellerman thinks that Bac. sorghi is able to penetrate the roots of the sorghum cane, as the young roots are often attacked during the disease, the infection coming apparently from the soil. Whether they pierce the epidermis itself, or enter by means of the root-hairs, he did not determine. Beyerinck found that through the root-hairs of the Leguminosæ, Bacillus radicicola was able to enter and cause the formation of tubercles.

The Action of Bacteria within the Tissues.

How are bacteria able to spread throughout the tissues of the plant? We have seen, in the results already detailed, that with certain forms, mainly saprophytic, they are able to pass from cell to cell. That they do this in plant-diseases is evidenced by the lesions that they call forth. But just *how* they are able to make their way from cell to cell is by no means so evident. In the light of our present knowledge concerning the transpiration stream, we cannot conceive of this current being utilized as a bearer of solid particles unless they have an inherent power of penetrating the cell wall. We do not find that the bacterial plant-diseases are able to spread their infective material throughout the plant in a manner comparable to a septicæmia, which is often developed in animals. This they would be able to do if the transpiration current could function, like the blood stream, as a distributor of infection.

It is possible that the lumina of the vascular tissue afford the least resistance to the spread of infection, yet, so far as we now know, only one parasitic species has adapted itself to this course. Wakker's Bac. hyacinthi affects primarily the xylem tissue of the fibrovascular bundle. Not only does it occupy the cavity of these air cells, but also attacks the surrounding walls, chiefly the middle lamella, which it soon converts into a disorganized gummy exudate, and is thus able to spread to the surrounding tissue.[1] Through these

---

[1] Wakker: Arch. néer., T. XXIII, p. 6 (1888).

elongated vessels it is able to spread the disease quite rapidly, as has been demonstrated by artificial infection experiments.

The rapidity with which the pear-blight germ is able to spread its infective material through a susceptible host also indicates a very rapid movement from cell to cell. This is especially marked in the softer succulent tissue of the youngest twigs and in the blossoms. After once securing an entrance to the rapidly growing tissues, it sets up a kind of fermentation which completely destroys many of the cells, thus forming large spaces which are filled with the gummy products of its fermentative activity. Under favorable conditions, the rapidity with which the blight bacteria spread is quite surprising. The following laboratory note may be considered fairly indicative of the rate of distribution. March 8, a Japan seedling was infected by puncture of the young stem. March 13, the disease had manifested itself by a local blackening of the tissue in the neighborhood of the inoculation point. Two days later, the stem showed that the disease had progressed fully six inches from point of inoculation, as indicated by blackened appearance. The presence of bacteria was also demonstrated microscopically fully an inch or more beyond this blackened tissue, showing that the spread of the disease, after having once established itself, was quite rapid.

It has been suggested that bacteria are able to pass from cell to cell through minute pores in the walls. Recent investigations[1] show that the direct union of the plasma of cell to cell is very much more widely diffused than was formerly supposed, and that *all* the *living* elements of the *whole* plant-structure of higher plants are thus united.

These plasmic strand-connections vary in diameter from $0.05–1.0\mu$, but on the average they are so small that it would seem hardly possible that they could be utilized by the bacteria in forcing their way from cell to cell. It seems much more probable that their progress is effected by ferment activity. In the case of the pear-blight germ and the hyacinth-disease, it is seen that the healthy tissue undergoes a decomposition under the influence of the bacteria, resulting in the production of a gummy substance, and in the case of the blight, the liberation of CO_2.

Bolley[2] finds the germ causing the surface scab in the potato im-

---

[1] Kienitz-Gerloff: Bot. Zeit. (1890), XLIX, 1.
[2] Bolley: Agric. Science, IV, 284.

bedded in the protoplasm of the cell. In this case the cell membranes were seen to be actually eroded by the bacillus.

A similar condition is found with B. oleæ-tuberculosis in the olive, and B. Veuillemini in the tumors of Pinus halapensis, where actual destruction of cell walls is accomplished under the influence of the germ. It is not at all improbable that those species which are adapted to a parasitic existence in the plant organism may not possess an eroding or fermentative ability, which would enable them to break down the resistant cell walls which impede their spread.

But how do we find it with those forms which are not so thoroughly adapted to this kind of life? It is much more difficult to determine the distribution in these cases than it is where the germ is able to grow luxuriantly. With the results which were obtained from the inoculation of those micro-organisms that are incapable of producing a genuine infection in plant-tissues, it has been shown in several cases that there was a distinct tendency to spread throughout the plant to a certain extent. These bacteria, which were determined at greater or lesser distances from the inoculation point, were also definitely located in the interior of the cells. This condition was best seen with cultures of B. acidi lactici, but was also recognized with B. luteus, B. pyocyaneus, and B. fluorescens. As has already been said, we cannot explain their presence unless they are able to pass through the cell walls. No openings of any kind could be determined, and the only remaining possibility that suggests itself is that they have the power, by means of a ferment excreted, to work their way from cell to cell without causing a permanent rupture. This we know to be the case with certain Ustilagineæ. They can penetrate the cell wall, which, after the passage of the hypha, again closes, so that no opening is apparent. With as small a structure as a bacillus this process is also conceivable. This explanation, however, does not rest upon experimental proof and is only suggested as a possible hypothesis.

RESISTANCE AND IMMUNITY OF PLANTS TOWARD BACTERIA.

The general exemption of plants from bacterial attack, which was referred to in the introduction under the expression "Immunity," reveals, upon a closer consideration of the question, a series of phenomena of a complex order. In the appendix to this paper will be found a complete list of all the plant-diseases which are now known to be closely associated with bacteria. A complete compilation of this sort has not previously been made for the bacterial diseases of plants, and a tabular review of this field of bacteriology was thought to be desirable. Although the causal relation between a specific organism and a plant-malady has not in every case been satisfactorily and thoroughly demonstrated, there is no doubt but that most of the plant-diseases mentioned may be rightfully ascribed to the ravages of these micro-parasites. When we consider how little attention has been paid to this branch of phytopathology, it is no wonder that our information on this subject is meager.

The list of diseases, although now limited, is rapidly and constantly increasing, so that we may freely predict that with a more thorough and exhaustive study of plant pathology from a bacteriological standpoint, the number of diseases will be materially augmented. Even now we possess sufficient data to qualify the assertion that plants are not subject to diseases of a bacterial origin.

A closer study of the general exemption of plant-structures from the attacks of micro-organisms reveals the fact that the phenomena heretofore embraced under the general term of immunity, are not of the same character in all cases. Different phases of this exemption seem to exist. One of these is the reaction of the plant toward micro-organisms in general. The other is the ability of certain plant-structures to withstand the inroads of a particular bacterial parasite. Under the first head, the micro-organism is unable to gain a foothold in the tissue of the plant, or, having once gained an entrance accidentally, it is unable to cope successfully with the repellant forces resident in the tissues. There is no *susceptibility* on the part of the plant toward the germ in question. This is the action which living tissue exerts in general. Where this action is overcome and the micro-organism triumphs, we have the development of disease.

Now, the ability of a single individual to withstand the attacks of a germ capable of producing a disease in the tissue of another individual of the same species, is evidently a different action. There is a certain degree of susceptibility on the part of the plant to succumb to this enemy, as is evidenced by the fact that when the conditions which maintain the natural balance of the forces inhibiting the germ are disturbed, the germ is then able to successfully attack the weakened plant. To this latter class of phenomena it would seem proper to limit the term immunity.

We would scarcely consider the human body immune from the attacks of ordinary saprophytes, even those forms which are normally found in the oral cavity. Most of them seem to possess no ability to thrive *inside* of the animal organism, but find their natural conditions of existence in the dead organic material which is always present in the mouth. Likewise, it would seem improper to say that a rose is immune from Bac. tuberculosis because the tubercle bacilli do not find in the tissues of the plant the necessary conditions for their development.

While the bacterial parasites already known are sufficient to indicate that we cannot consider the vegetable kingdom as wholly free from bacteria, yet we must admit that the susceptibility of plants is very much less than that of animals.

In considering, then, this exemption of plants from the attacks of micro-organisms, we will divide the phenomena into two classes:

First, those due to what may be called *Resistance;* second, those due to *Immunity.*

Before going farther, it will be necessary for us to determine the limitations which will be imposed upon the meaning of these two terms.

The inherent power of the vegetable organism to withstand the action of bacteria in general may be termed *Resistance.* This resistance which the plant offers to the entrance of micro-organisms may be due to various causes, and is operative throughout the whole range of plant life. It is the normal condition of the plant, and is closely correlated with the conditions of nutrition, for when the natural play of these forces is disturbed, and an abnormal state of affairs supervenes, this power of resistance may be subject to greater or less modification. This lowering of the general vitality of the plant, due possibly to a number of causes, is usually manifested in a lessening of the powers of resistance which the plant seems to possess. The plant organism becomes then more susceptible to the attacks of disease.

This state of affairs must not be confounded with a condition which affords a more favorable opportunity for the development of the attacking parasite. The one concerns itself with those processes which tend to lower the general vitality, and thus the resistance of the plant; the other relates only to those conditions which give increased powers of development to the attacking organism. It is possible that the same set of causes may produce this double effect, as, for instance, such meteorological conditions as excessive moisture, to the extent that it not only interferes with the normal action of the vital processes of the plant, but gives also the optimum conditions for the development of the parasite. But while the resultant of these forces may have a doubly deleterious action on the plant, it does not of necessity follow that this should be the case.

It will have been seen that the resistance which a plant offers toward its enemies in general is broad and wide-reaching in its effects; not so with immunity in regard to a specific disease. The one is a general condition of normal, healthy vegetable life; the other, the expression of a restricted group of vegetable organisms toward the cause of a specific malady. The term immunity will then be restricted to those plant-organisms which do not succumb to the action of a germ that is able to call forth a genuine infection in another related species. Thus we may define *Immunity* in plants as the ability of the organism to repel the attacks of a germ which produces a pathological condition in a closely allied form. By a closely allied form, we mean a species or variety which stands in close taxonomic affinity with the form which is a natural host for the germ in question. Now, the limit of this immunity must, of necessity, be a somewhat variable one. Whether the term should be restricted in its application to those species grouped in the same genus or subgenus, or whether it should embrace the limits of a whole family, will differ in different cases.

Our knowledge of the ability of bacterial germs to produce in plants a diseased condition in different species is yet somewhat limited, but the observations already on record are further strengthened by analogous cases with fungal diseases.

We know that some fungi are very restricted in their development, and that outside of a single host-plant they are unable to call forth any diseased condition, even under the most favorable opportunities. This is the case with *Fusicladium pirinum*, which causes the destructive pear-scab both in this country and Europe. Sorauer[1] claims that under no conditions, even during the seasons which are most favorable for the growth of the fungus, does it ever exceed the

[1] Sorauer: Landw. Versuchsstat. XXVI, 327.

narrow limits which form apparently its fixed boundaries. On the other hand, the host-distribution of some diseases is known to be dependent largely upon certain climatic conditions. During years with the normal amount of rainfall or seasons of drought, the fungi only attack the ordinary hosts upon which it is a natural parasite, but when an excessively rainy season supervenes, the conditions being then much more favorable for the development of the parasite, the fungus is reported upon many new hosts, usually, however, allied species within generic or tribal limits. Swingle[1] gives an interesting series of results which he found with the Peronosporeæ on the Euphorbiæ in Kansas, corroborative of this statement. Here the ordinary boundaries which usually limit the spread of the fungus were broken over during those years that were particularly favorable for the development of the parasite, and the same parasitic species was often found on entirely new hosts.

These two extremes indicate the impossibility of establishing any hard and fast line for the limits of a disease, and, consequently, the limits of immunity in the latter example would be much wider than in the former.

So far as is at present known, among the bacterial plant-diseases we have no malady that is able, either in a state of nature or by artificial injection, to produce a pathological condition in plants belonging to different natural families. Such a condition may not be impossible, however, and further research may give us examples which have a wider range of host-plants. The majority of them are naturally limited in their distribution, even within generic and often within specific boundaries.

The two phases to which this exemption of plants from bacteria is due, immunity and resistance, although acting distinctly with reference to different germs, may be resident even in the same plant organism. A plant may be resistant or totally insusceptible toward an ordinary saprophyte or even a bacterial parasite whose host-plant is in a distant family, and yet may or may not possess immunity from another species of bacteria which is a natural parasite upon a closely related species.

The presentation of a few examples of what is meant by this will suffice to illustrate this distinction between immunity and resistance.

[1] Swingle : Kans. Acad. of Science, Vol. VI, 1887–88.

Reference must again be made to the pear-blight germ (B. amylo-vorus), as its biology is the most thoroughly investigated of any of the bacterial plant-diseases. So far as is known, this malady is con-fined strictly to the Rosaceæ, and almost without exception to the sub-order Pomeæ. It has been reported to have been found in the young fruit of the Kelsey plum[1] (Prunus sp.), where it was traced to the sting of a curculio, when the eggs were deposited, and also in a new disease of rasp- and blackberry.[2] However, this last exception has not yet been thoroughly demonstrated. Arthur also reports that he was able to produce a slight infection in the succulent shoots of the peach, but not at all in other non-rosaceous fruits, as mulberry and grape.[3] While the disease afflicts several different species of Pomeæ, its commonest host is the cultivated pear. The early history of this disease shows that it was at first more or less restricted in its development on this species, attacking only certain varieties, but in the wider range of the malady in later years, it seems to have acquired the ability of successfully attacking other varieties, until now we know of no variety that is absolutely immune, in the state of nature, from the disease. Although not wholly immune, many horticultural strains possess immunity in a partial degree, as is evi-denced by the fact that under like conditions certain varieties yield much more readily to the disease than others. By some peculiarity in the structure of the plant, possibly merely mechanical in its nature, one variety is able to successfully resist the attacks of the parasite to a larger extent, and thus possesses a partial natural immunity from the disease.

This immunity can often be overcome, however, if the germs are able to gain access to the tissues in some manner, as by wounds from insect stings, etc.

This is well shown by the following greenhouse experiment:

Set 1. Two pear trees (Japan seedlings[4]), 2 years old, were inocu-

---

[1] Personal communication from Mr. M. B. Waite, to whom I am indebted for a number of facts bearing on this topic.

[2] Detmers: Ohio Bull. Exp. Stat., No. 6, Oct. 1891.

[3] Arthur: N. Y. Ann. Agri. Exp. Stat., 1884, 362.

[4] This stock, lately introduced from Japan, is in great favor with nurserymen on account of its vigorous and luxuriant growth, and its seeming refractory qualities, under ordinary conditions, toward the blight. Growers have, however, not had experience with it long enough to determine whether its seeming good qualities are of a permanent nature or not.

lated with B. amylovorus, sub-epidermally, on March 3. March 12, the blight was well marked as a blackened patch for nearly an inch on each side of point of inoculation. Period of incubation, therefore, 9 days.

Set 2. Two 2-year old trees, of blight-proof variety, budded on Japan stock, were inoculated, sub-epidermally, March 17. Evidence of blighting in leaves and stems readily recognizable on March 25. Incubation period, 8 days.

Set 3. 2-year old budded Clapp's Favorite (a variety readily susceptible to the disease in the orchard) was inoculated in the same manner, March 3. March 12, diseased condition apparent, although not as well marked as in Set 1. Period of incubation, 9 days.

From this set of experiments, it will be noted that those varieties (Sets 1 and 2) which under natural conditions are known to be much more refractory than others (3), not only lost their partial natural immunity when subjected to artificial inoculation, but yielded to the disease fully as quickly as did the variety that was naturally susceptible. This artificial inoculation is, however, a much severer test than they receive under the operation of ordinary conditions of cultivation, but it shows that the varying degrees of susceptibility, or, in other words, immunity can be overcome under certain conditions.

Waite attempted to infect two different species of Cratægus (C. oxyacantha and C. tomentosa, var. parviflora) by atomizing the flowers, but found that only the latter succumbed to the action of the germ. Here again is a case of immunity where one variety is exempt, under certain conditions, from the disease. The immunity in this case is, however, not a deep-seated condition, as is shown by the successful infection of the latter variety by *puncture* inoculation, where the bacteria were actually introduced into the tissues. Other species of the pomaceous group of the Rosaceæ are more or less susceptible to the disease, although much less so than the pear and apple. The cultivated species of this fruit family seem to be more readily susceptible than many of the wild species. Certain species, such as the Haw (Cratægus spp.) and Shadbush (Amelanchier Canadensis), are subject to the disease when artificially inoculated with the germ.

The researches of Savastano[1] on the tubercle of the olive tree in Italy also aptly illustrate the relation of immunity and resistance.

[1] Savastano: Tuberculosi, iperplasie, e tumori dell' olivo: Ann. R. Scuola Sup. d'Agric., Vol. V, 1887.

The bacillus causing this disease produces an hypertrophy of the cambial and extra-cambial tissue, which is, however, quite localized. The results of artificial inoculation into young olive trees were evident in 25 days, and in 2 months a 'well-marked swelling was to be noted at seat of inoculation. Savastano was not able, however, to successfully infect, in equal degree, all varieties of this species, but found a varying degree of susceptibility in different varieties. When the B. oleæ-tuberculosis was injected into other healthy fruit-bearing trees, such as lemon, bitter orange, pear, apple, quince, etc., no trace of the infection could be noticed. Here is, then, a case of resistance against the organism even though it was a parasitic organism on other forms of plant life.

He also infected olive trees in like manner with several other micro-organisms which he found associated in the tuberculous growths with the true cause of the disease, but in no case was any pathological condition brought about, showing that the normal resistance of the plant was able to overcome the accompanying saprophytic micro-organisms, although it was not able to withstand the attacks of the parasite in all cases.

The examples already cited will suffice to show the distinction made between immunity and the normal resistance of the plant, and also that immunity is a varying term itself, sometimes applicable within narrow limits, then again spreading over a greater variety of species. Not only are the limits of immunity ill-defined and varying, but the degree of immunity also varies considerably. This is readily recognized in horticulture, when we say that one variety is more susceptible to the attacks of a certain disease than another. This variation of the susceptibility of different varieties indicates that they possess an immunity to a greater or less extent from the action of a definite specific germ.

We might consider, also, other phases of immunity which present themselves for consideration, but a mere mention of these will be all that can be given of them in this connection. Thus we have the *local immunity,* which certain tissues enjoy against the attacks of micro-parasites as they increase in age and consequently become better developed and more resistant. This rests on a purely physical basis, and the importance of it will be considered later more in detail. This local immunity of certain tissues is strongly reinforced, also, by a consideration of the phenomena connected with certain fungal diseases, as in the case of Cystopus candidus, which, according to De Bary, can only successfully infect the host-plant when it enters the young cotyledons of the plant. Hypoderma macrosporon is only able to gain entrance into the pine through the young pine cones.

This law usually holds in connection with the bacterial plant-diseases. Most of them are more virulent in their course when they attack young and undeveloped hosts, and not a few are apparently inhibited after the tissues have reached a certain stage of maturity, as in the case of the oat-disease of Galloway and the pear-blight.

Causes of Immunity and Resistance.

Having cited special cases illustrative of resistance and immunity, we may now turn to the consideration of the possible factors which are able to cause these conditions. To this difficult problem we cannot hope as yet to give any definite and conclusive answer. The rapid progress which has been made in this department of animal biology within the past five years indicates how vast and complicated the question is in all its bearings. The accumulating data which have already been collected have, however, contributed much to a better conception of the problem of immunity, and lead us to believe that any experimental consideration of this subject in relation to plants, even though negative in its results, is not entirely without value.

The attempt has been made in the previous pages to show that there are two factors at work in the struggle of the plant with its parasitic enemies. We shall not, however, attempt to prove that the phenomena which are considered as resistance and immunity are brought about through the action of separate and distinct forces. It is possible, and quite probable, that the conditions which produce one factor may also be the cause of the other. For instance, in animal biology, Wyssokowitsch determined that certain saprophytes, as well as some parasites, disappeared completely when introduced into the blood of a living rabbit. Nuttall determined that the cause which destroyed saprophytic as well as pathogenic organisms was the same, namely, the germicidal property of the body fluids. Although the result was brought about through the operation of the same cause, we would not consider these two cases similar. The one is simply the normal resistance which the animal offers to an organism which has no power under any circumstances to produce a diseased condition in its body. The other is a case of immunity to a certain degree against the pathogenic organism.

We cannot, however, compare the phenomena of animal immunity too closely with that of plants, as we find that the causes which tend to produce this state are unlike in the two kingdoms. A much closer similarity exists between the immunity of plants from bacteria and from fungi. Not only are the phenomena presented quite homologous in character, but the causes which are operative are no doubt more or less closely related.

It will be hardly possible to classify the causes which may tend to produce the resistance and immunity of the plant from its bacterial foes in any satisfactory and complete manner. A tentative classification, however, may be suggested along the lines of physical and chemical action ; including under physical sources all those mechanical contrivances, such as epidermal covering, cutinization of tissues, and the secondary thickening and lignification of cell membranes, which enable the plant to ward off injurious external forces. Under chemical sources we would class, not only the reaction of the tissues and the nutritive conditions, but the resistant action due to the living protoplasm itself.

That one of the possible causes of the partial immunity which some plants exercise toward parasitic forms is dependent largely upon the mechanical obstructions which the plant offers to the entrance of the germ, is seen in the examples of immunity which have already been mentioned.

Certain strains of the common pear, under ordinary conditions of cultivation, are quite refractory toward the blight when subjected to natural infection which takes place in the blossom through insect visitation, but these varieties when artificially inoculated sub-epidermally with bacteria yield readily to the disease. This can scarcely be accounted for on any other ground than that the blight bacteria are unable to gain a foothold on account of some peculiarity of the external plant membranes. This may be so slight that one cannot detect any histological difference in the exterior cells, yet one variety will be considerably more refractory under natural conditions than the other.

Arthur[1] has drawn attention to this point and suggests that these differing degrees of susceptibility are due to physical causes.

Not only has mechanical exclusion, as a cause of immunity, been shown in the pear-blight disease, but other bacterial diseases, as well, illustrate the operation of this cause. In his inoculation experiments upon "La Jaune des Jacinths," Wakker[2] infected, besides a number of susceptible varieties of the hyacinth, one variety (Norma) which was regarded under natural conditions as immune from the disease. This he found succumbed as readily to the malady as the naturally susceptible varieties when the infective material was introduced into the leaf of the plant, showing that the natural immunity of the variety was dependent upon its epidermal tissue.

These two cases, cited above, are examples where immunity of the plant toward a specific parasite is dependent upon a physical means of exclusion, but the same cause is also operative in regard to the general resistance of the plant against all forms of germs. Under ordinary circumstances we do not find that saprophytic organisms, even decomposition bacteria, are able to enter the normal, healthy, intact plant structure, yet we have shown in the preceding pages that these germs when they once get a foothold into the plant tissue are able to survive for a long time, and where the vitality of the host is much reduced may even cause a disorganization of tissue.

If the thin cutinized layers of epidermal cells afford such an effectual barrier to the entrance of micro-organisms, the more resistant corky layers of the mature plant will be even more efficacious in excluding germ life in general. This is demonstrated by the complete inability of most bacterial diseases[3] to penetrate the resistant tissues of the bark. These act as an efficient barrier against the entrance of any organisms, except through the natural openings (lenticels) and wounds.

Although the cuticular and cortical layers of the plant function as the chief hindrances to the entrance of germs, fungal as well as bacterial, the fully developed walls of the inner cells also inhibit the spread of pathogenic forms. The young fruit of the pear cannot be successfully infected after it has reached a certain size, as

[1] Arthur: Proc. Phil. Ac. Nat. Sc. 1886, p. 340.
[2] Wakker: Arch. néerland. d. Sc. ex., T. XXIII, p. 18.
[3] Savastano thinks that B. oleæ-tuberculosis must make its way through the bark tissue in order to reach the succulent cambium where it thrives. But this idea is conjectural and is not based upon experimental proof.

the tissues become so mature that the germ is not able to pass from cell to cell. It is also particularly prominent in the blighting stem, where the disease is usually confined to the youngest, most succulent tissue. This is characteristic of bacterial plant-diseases in general, that the most pronounced lesions are always found in tissue which has not yet lost its power of growth.[1]

We turn now to consider the chemical sources of protection which the plant may possess against its enemies. First of all we have the chemical reaction of the plant-tissues. As was noted in the introduction, it is to this source that most authors ascribe the general freedom of plants from bacteria. This was before it was thoroughly proven that vegetable structures were affected to any extent by bacterial diseases, and was probably based upon the then prevailing idea that bacteria required alkaline and fungi acid substrata for their development. So many exceptions to this law are now known that this statement has lost much of its original force. Most of those forms which we know to be able to cause bacterial plant-maladies are usually more or less indifferent to the reaction of the medium, growing in either weakly acid or alkaline nutrient media. Arthur succeeded in raising B. amylovorus in 2 per cent malic-acid bouillon.

The experiments already detailed indicate that saprophytes, and even some animal pathogenic forms, are not destroyed by the plant juices either within or outside of the plant. Forms like B. prodigiosus, B. lac. aerogenes, B. megaterium, B. coli commune, Kiel water-bacillus and blue pus germ grew in the expressed plant juices of various kinds and produced a considerable turbidity in 24–48 hours.

It would seem, then, that the importance of this factor as the prime cause of immunity and resistance has been considerably overestimated. Very much more stress is to be laid upon the mechanical barriers which the plant possesses against the entrance of germs, as well as the activity of the living protoplasm.

The nutritive conditions offered by the plant may also be considered in this connection. While the dead tissues of many plants

[1] The observation which Prillieux noted in the case of B. Vuillemini, which causes the tumors in the *old* wood of the Aleppo pine (Pinus halapensis), is only apparently contradictory. The fact that a local hypertrophy of tissue was produced by means of the bacterial stimulus showed that the tissues were yet in a secondary meristematic condition.

[2] Zopf: Die Pilze, S. 173.

afford a tolerably good substratum for the development of many species of bacteria, and thus show that the tissues are not wanting in the necessary nutritive materials, the conditions are different in the struggle between the disease germ and the living plant. Unless the germ is able to gain access to the inner tissues by means of accidental lesions, it must force its way through the epidermal cell walls, or the cell membranes of the interior tissue, either directly from the outside, or after having first gained an entrance through the stomata into the intercellular spaces. In either case it has no supply of nutrient material with which to carry on its metabolic activity, and is therefore unable to gain a foothold from which to develop.

If the bacteria could increase in sufficient numbers on the surface of the plant they might be able to penetrate the tissues more easily. That B. amylovorus is in this way able to gain access to the tender tissues of the receptacle of the pear-flower is extremely probable. The sugary secretions which are poured out by the nectariferous glands in the flower induce insect visitation, and afford the best possible medium for the growth of the bacilli which are brought to it in the sticky exudate which adheres to the insect. Thus the bacteria have a fertile soil prepared for their further development, and are able to multiply with ease. Having thus secured such a coigne of vantage, they are able to penetrate the non-cutinized tissues beneath, and thus gain entrance to the inner parts of the plant.

Do plant juices possess germicidal properties?

Recent experiments in animal pathology have demonstrated that the blood of many animals possesses the property of destroying bacteria to a limited extent, either when brought directly in contact with it in the body of the animal, or after it had been aseptically removed. This bactericidal power of the blood enables the animal to destroy, not only the saprophytic forms with which it comes in contact, but also a certain number of even malignant bacteria, which otherwise would be able to multiply in the body and finally produce death. This line of investigation, so fruitful of results in immunity in the animal kingdom, suggested a series of experiments as to whether a corresponding condition might not exist in plant life as well.

Although we have in plant-tissue nothing homologous to the circulatory fluids of the animal body, it might be conceivable that the

plant fluids were endowed with a germicidal property for the protection of the plant. If the plant possessed such a peculiarity, an
examination of the cell sap, with this point in view, ought to indicate
its presence. The first difficulty to be met with, however, is to
secure the cell sap free from solid elements and in an aseptic condition. This can be accomplished without serious difficulty in the
case of the blood, but as there is no special channel for the movement of the plant fluids, the problem here is not so easily solved.

In a foregoing set of experiments, in which different species of
bacteria were artificially introduced into the plant-tissues, it was found
in a number of cases, particularly with saprophytic forms, that they
had multiplied to a limited extent. This at first might indicate that
the plant-juices had no germ-destroying power. Such a conclusion
need not necessarily follow. It has already been found that the
germicidal action of the body fluids towards certain organisms has
a definite limit, and that when too many germs are brought in contact with it its capacity for destroying the bacteria is overcome,
and the germs are able to increase and call forth a pathological condition in the body. That such might have been the case in the
above experiments was possible, as large numbers of bacteria were
introduced into the plant.

To determine this, experiments were carried out on plant-tissues
to determine if the plant-juices possessed any germicidal properties.
Heating could not be resorted to as a means of sterilization, as this
would affect any property analogous to the germ-killing peculiarity
of the blood, so the only recourse was to obtain the juices aseptically.
Trituration in a small sterilized mortar was first attempted, but this
method was regarded as unsatisfactory on account of the cellular
detritus present. Expression of the juice was then tried by com-
pression. A small screw press, capable of being sterilized, was used
for this purpose, and as the fluid escaped through the minute openings in the bottom it was caught in a sterilized receiver and pipetted
into small culture bulbs as before. By this process the plant-juice
was secured perfectly free from solid particles. Into this culture
fluid a number of germs were inoculated, as determined by Nuttall's[1]
device for accurate quantitative work, and at varying intervals of
time the germ-contents of the bulbs were determined. In each case

---

[1] Nuttall: Bull. J. H. H. No. 13, May-June, 1891.

about 5–7 cc. of the expressed juice was used. The quantitative results, indicated in the following table, show the influence of the plant juices upon the growth of different species of micro-organisms.

Name of Germ.	No. used as "Seed."	Period of Incubation.	No. at end of Experiment.	Plant Juice used.
Kiel water-bacillus ..	859	2 hrs. 40 min.	6700	Canna.
	26	24 "	8420	"
Bac. lactis aerogenes.	26	5 "	90	"
	26	24 "	7200	"
B. coli commune.....	46	24 "	12,420	"
B. megaterium	14	1 hr. 45 min.	25	"
	14	7 "	215	"
	14	10 hrs. 30 min.	3060	"
B. prodigiosus.......	20	4 "	35	"
	20	24 "	4600	"
	105	4 "	160	"
	105	24 "	32,000	"

Most of the species which were used in the above experiment are those which showed a marked increase when in the plant-tissues for a considerable period of time. This table, however, indicates that the increase began immediately upon the introduction of even small numbers into the cell sap, showing that this fluid possessed no bactericidal properties.

The attempt to test the action of cell sap was also tried in another way. However, this method did not allow of accurate quantitative determination, although the other conditions were more nearly normal. It was based upon the action of root pressure in the plant. Thrifty young plants with good-sized stems, like the Lima bean or geranium, were selected, and after having washed the stems with a disinfecting solution, they were cut off with a sterile knife about an inch from the ground. The stump of the plant was then quickly covered with a short sterile test-tube, which made a moist chamber that prevented evaporation. The pot was then set in a warm place to induce copious root-action, and in 12 to 24 hours a large drop of cell sap had exuded from the cut end of the stem. This clear fluid was then inoculated with a few germs from a fresh culture, to be tested, and after a varying length of time their relative number determined as nearly as possible.

A marked increase with B. megaterium, B. butyricus, B. coli commune and B. pyocyaneus was noted, while Strept. pyogenes in five days was killed. As the death of the bacteria inoculated did not occur at once, but was gradual, it was no doubt due to unfavorable nutritive conditions rather than any germicidal effect of the cell fluids.

The cell sap usually possesses a distinct acid reaction, and would, no doubt, inhibit the growth or kill out by malnutrition those forms susceptible to acid reaction. Some plant-juices afford a much better nutritive medium than others, such as the sugar-cane or the saccharine varieties of sorghum, or the sap of such trees as Acer saccharinum. Sternberg[1] has recommended the milk in green cocoanuts as a nutritive medium for even animal pathogenic organisms. This is really the elaborated cell sap of the embryo-sac.

In the animal body we find that a bactericidal property is not only resident in the blood plasma and tissue juices, but is also found in various secretions and excretions which are formed in the animal organism.

Thus the sputum of a healthy individual is known to have an anti-bacterial effect on anthrax,[2] while the germicidal properties of fresh milk[3] and urine[4] are quite considerable.

Although we have been unable to detect any analogous property in the cell sap of the plant, we know that the plant is able in many cases to protect itself by means of its secretions. Thus conifers are protected from disease in many cases by the copious flow of turpentine, which forms an effectual barrier against the entrance of fungi as well as bacteria.[5] The ethereal oils which are found so widely distributed throughout the plant kingdom are known to possess the ability of hindering, and in many cases actually destroying, germs when brought in contact with them.[6] In all probability they function in a similar way in the plant, although many parasites may have adapted themselves to this condition.

---

[1] Sternberg : Phil. Med. News, 1890, p. 262.

[2] Nuttall : Boylston Essay, Harvard, 1888.

[3] Fokker : Zeit. f. Hyg., IX (1890), 41.

[4] Lehmann : Cent. f. Bakt. VII (1890), 457.

[5] Hartig : Die Baumkrankheiten, S. 139, 166.

[6] Cadeac et Meunier : Ann. de l'Inst. Past. 1889, 317. Freudenreich : Ann. de Microg., 1889.

Besides the possible sources of immunity and resistance which we have already considered, we have the action of the living plasma of the plant. Concerning the exact nature of this force we have but little positive knowledge, although it is probably chemical in its action. We know that the plant endowed with vital activity is more resistant toward outside influences than the same dead structure; also, that protoplasm which is in an active state is much less subject to the attacks of disease than quiescent or inactive protoplasm. This has been demonstrated in the case of a number of tree-destroying fungi that are only able to overcome the tissue cells during the winter season of rest.[1] Anything that tends to impair the normal exercise of the vital functions of the protoplasm predisposes the plant to the attack of outside organisms. It is quite unlikely that this so-called lowering of the general vitality affects to any considerable extent the physical means of resistance. It is much more probable that it is the repelling ability of the living protoplasm that is weakened, and thus less resistance is offered to disease. To this action Marshall Ward attributes a large share of the resistance of the plant against its parasitic foes.

Speaking of fungi, he says, so long as the protoplasm can over-come, by respiratory oxidation or otherwise, the small amounts of poison generated by the parasite, the hypha does not pass, but when the poison exceeds this power of repulsion, then it effects an entrance into the cell.[2]

In here presenting the sources which seem to be operative in the production of the resistance and immunity of plant-tissues, examples have also been given illustrative of this condition with fungi as well as bacteria. While the conditions necessary for the best develop-ment of these two classes of vegetable life are often different, there can be but little question that the same means of protection which the plant possesses, operates in many cases against the one quite as effectually as against the other. The refractoriness of higher plants against bacteria has many more points in common with the same phenomena against fungi, than it has with the action of the animal body against bacterial life.

[1] Hartig: Die Baumkrankheiten, S. 87, 112, 33 u. A.
[2] H. Marshall Ward : Journ. Roy. Soc., 1890, 213.

CONCLUSIONS.

For sake of convenience, a short review of the points which have been developed in the preceding pages will now be made.

1. The increasing importance of the bacterial portion of phytopathology necessitates a more thorough investigation of the influence of bacterial life in general upon plant-tissue than has heretofore been considered necessary.

2. The artificial inoculation of higher plants with different microorganisms (not known to be pathogenic for plants) reveals the fact, contrary to the usually accepted idea, that quite a goodly number of different species are able to withstand the action of the living plant organism for a not inconsiderable length of time.

3a. Of the species which are able to live in plant-tissues for a considerable period of time, those which are ordinarily adapted to a saprophytic method of existence are particularly prominent. Not all saprophytes, however, possessed this power, but in certain forms, as B. fluorescens, B. acid. lact., B. butyricus, etc., it was a marked feature.

3b. Among those forms which are facultative parasites upon the animal body, but few were found that seemed to be able to live in plant-tissue. With the exception of B. pyocyaneus and the Schweineseuche bacillus, they gradually decreased in numbers and finally died.

3c. The inoculation of plants, not taxonomically related to the natural hosts of bacterial plant parasites, with species of microorganisms naturally parasitic on vegetable tissue, showed that while the bacteria were unable to spread, they could survive at the inoculation point in large numbers.

4. Not only were numbers of different species of bacteria able to *live* in the plant from 40 to 80 days or more, but many of them (mostly saprophytes) were able to *spread* throughout the tissue of the plant to a limited extent (20 to 50 mm. or more).

5. The local distribution always took place in an upward direction, and the bacteria were found to be generally *intracellular* instead of *intercellular*.

6. According to the present views of physiologists regarding the transpiration stream, it does not seem possible that this current can

account for the distribution of bacteria in the tissue. It seems to be correlated much more closely with the actual *growth* of the micro-organisms.

7. The facts already determined relative to the ability of sapro-phytes to thrive in plant-tissues throw some light on the question of the normal presence of bacteria in healthy plants. A large number of cultures made from the inner tissue of healthy plant stems revealed no bacteria, but where stems were wounded, even by a needle-prick, the bacteria on the surface were able to enter and live for a long time. Thus it is possible that bacteria may enter through lesions so small as to escape notice, or they might even live in the tissue after the wound had healed over.

8. With bacteria, not adapted for growth in plants, I have been unable to prove that they could enter the plant where the epidermal tissue was known to be intact. In the case of parasitic species on plants, they sometimes effect an entrance into tissues without the intervention of wounds of any sort.

9. The actual method by which saprophytes are able to spread in plant tissues has not been satisfactorily determined, but it seems that the cellulose wall undergoes a change that renders it permeable to the bacteria. Those forms causing a pathological condition in the plant spread, in many cases, by means of the fermentative and destructive power they possess.

10. The phenomena, heretofore regarded as immunity of plants from micro-organisms, present two phases so distinct in their action that it seems proper to separate them to a certain degree. The exemption of plants from bacteria in general is due to what may be termed the *resistance* of the plant, while the more restricted term, *immunity,* is reserved for the ability of a certain group of plants to be refractory toward a disease germ that is able to cause a patho-logical condition in closely allied forms of plant life. No hard and fast limits can be drawn for the immunity of plants, as this condition varies in each disease. The causes which bring about this ability of the plant to repel not only bacteria in general, but those toward which it is somewhat susceptible, are various.

In the case of immunity, physical causes, such as the epidermal and cortical resistant tissues, matured and thickened cell walls of the inner tissue, exclusion by gummy exudates, etc., are the leading factors. The exemption of plants from bacterial diseases, however, does not rest upon any single factor but upon the interaction of various causes.

Added to this mechanical source of immunity are the chemical reaction of the juices, the unfavorable conditions of nutrition, the action of the living protoplasm, etc., all of which exercise an unfavorable or inhibitory effect on bacterial life.

The whole question of immunity of plants from bacteria is much more closely related to the same question as regards fungi than it is to the subject of immunity as seen in the animal kingdom. Vegetable cell juices, aside from their acid reaction, are entirely powerless against bacteria, and do not possess any germicidal properties like the blood-serum of animals.

BIBLIOGRAPHY.

A. *On the normal presence of bacteria in healthy tissue:*
1. Bernheim : Münch. med. Wochen., 1888, s. 743.
2. Buchner: Münch. med. Wochen., 1888, No. 52.
3. de Vestea: Ann. de l'Inst. Pasteur, 1888, 670.
4. Fazio: Revista inter. d'Igiene, 1890. (Abs.: Cent. f. Bakt. VII, 798.)
5. Fernbach: Ann. Past., 1888, 567.
6. Groucher et Deschamps: Arch. Med. Exp., 1889, 53.
7. Galippe: C. R. Soc. Biol., 1887, No. 25.
8. Laurent: Bull. l'Acad. roy. de Belg., t. X, 38.
9. Laurent: Bull. l'Acad. roy. de Belg., t. XIX (1890), 468.
10. Lehmann: Münch. med. Wochen., 1889, No. 7.
11. Ralph: Trans. Royal Soc. Victoria, Vol. XX, 1884.
12. Van Tieghem: Bull. Soc. Bot. de France, 1884, XXXI, 283.

B. *On the artificial inoculation of plants with bacteria, non-parasitic in vegetable tissue:*
1. Lominsky: On the parasitism of some pathogenic microbes for animals in living plants. Wratsch, 1890, No. 6. (Ref. Cent. f. Bakt. VIII, 325.)
2. Savastano: Tuberculosi dell' olivo: Ann. R. Scuola. Sup. d'Agric. in Portici, Vol. V, 1887.

APPENDIX GIVING A LIST OF THE BACTERIAL PLANT DISEASES, WITH BRIEF RÉSUMÉ OF THEIR PRINCIPAL CHARACTERISTICS.

The preparation of a complete abstract of the bacterial diseases of plants at present is attended with some difficulty. Besides the number of well authenticated and confirmed observations upon this class of diseases, there are quite a number of maladies which have been as yet incompletely worked out. The disease has been experimentally reproduced only by inoculation of diseased tissue, not by infection from a pure culture of the germ. The classic canons of Koch, which are regarded as essential in the elucidation of the etiology of an animal disease, are, however, just as applicable in the investigation of plant maladies, and we can consider no disease as sufficiently proven to be of bacterial origin until the germ has first been isolated, and then successful inoculation experiments made with the pure culture of the organism.

The literature embracing this branch of phytopathology is largely American, but it is widely scattered, and it was thought that a brief résumé of the bacterial plant-diseases known to date would be of value. So far as I am aware, this has not yet been attempted in any complete degree. Comes' Crittogamia Agraria (Naples, 1891) and Ludwig's Lehrbuch der Niederen Kryptogamen (Stuttgart, 1892) are the only works from a European source that attempt to deal in any satisfactory way with the bacterial plant-diseases, and even these include only a part of the diseases already known.

Tables I and II give a list of those diseases which have been traced to a bacterial origin and where the disease is claimed to have been experimentally produced by inoculation of pure cultures. Table III gives a provisional list of diseases possibly of bacterial origin, but the etiology of each malady has not yet been traced to a specific microbe, and consequently cannot be regarded as thoroughly proven.

The number of plant maladies caused by bacteria might have been considerably increased if all the cases on record of the presence of bacteria in diseased tissue had been added. This, however, does not signify that the bacteria bear any etiological relation to the disease, and it is possible that in many of these cases they are present only as decomposition-organisms in dead or dying tissue. A considerable number, however, of these have been added, as it has been suggested that it might be useful to collect the data on this subject, imperfect or otherwise, which at present are so widely scattered in various publications.